Disposal
of clinical waste

*HEALTH SERVICES
ADVISORY COMMITTEE*

HSE BOOKS

Safe disposal of clinical waste

© Crown copyright 1999

Applications for reproduction should be made in writing to:
Copyright Unit, Her Majesty's Stationery Office,
St Clements House, 2-16 Colegate, Norwich NR3 1BQ

First published 1992
Second edition 1999

ISBN 0 7176 2492 7

All rights reserved. No part of this publication may be reproduced, stored in a retrieval system, or transmitted in any form or by any means (electronic, mechanical, photocopying, recording or otherwise) without the prior written permission of the copyright owner.

This guidance is prepared, jointly by HSE and the Environment Agency, through a Working Group of the Health Services Advisory Committee, appointed by the Health and Safety Commission as part of its formal advisory structures. It has also been produced with the agreement of the Scottish Environment Protection Agency. The guidance represents what is considered to be good practice by the members of the Committee and has been agreed by the Commission. Following this guidance is not compulsory, and you are free to take other action. But if you do follow this guidance you will normally be doing enough to comply with the law. Health and safety inspectors seek to secure compliance with the law and may refer to this guidance as illustrating good practice.

Contents

Introduction ...1

Who should use this guidance? ...2

Definition of clinical waste ...3
Treated waste ...4
Special waste ...5
Waste from human hygiene ...5
Human tissue ...7
Definition of clinical waste for transport ...7

Managing the risks to people - the law ...9
The Control of Substances Hazardous to Health Regulations 1999 ...10
The Management of Health and Safety at Work Regulations 1992 ...10
Consulting employees ...11
The Transport of Dangerous Goods (Safety Advisers) Regulations 1999 ...11

Managing the risks to the environment - the law ...13
Duty of care ...13
Other legal requirements ...15
Special waste ...15

Precautions for dealing with clinical waste ...17
Training and information ...18
Personal hygiene ...20
Personal protection ...20
Immunisation ...21
Segregation ...21
Handling clinical waste before disposal ...23
Storage ...25
Packaging ...26
Labelling ...29
Transport ...30
Dealing with accidents, incidents and spillages ...32

Treatment and disposal ...34
Treatment methods ...34
Disposal methods ...35
Summary ...40

Additional advice for producers other than hospitals ...42
Clinical waste from general medical practitioners ...43
Clinical waste arising from home treatment ...43
Clinical waste from veterinary centres and practices ...43
Clinical waste arising from research ...44
Clinical waste from ambulance services ...44

Monitoring and review ...46
Investigation of accidents and incidents ...46
Monitoring results ...47

Appendices
1. Classification of and packaging requirements for clinical waste ...48
2. First aid for sharps injuries ...50
3. Example of a local clinical waste disposal procedure ...52
4. Example of a local mercury spillage procedure ...58
5. Management checklist ...60

References ...61

Glossary ...65

Introduction

1   Large quantities of clinical waste are produced every day from a whole range of workplaces. Unless the segregation, handling, transport and disposal are properly managed, such waste can present risks to the health and safety of people at work, members of the public, and the environment.

2   Different laws cover the risks to people and to the environment from clinical waste. These laws are enforced by different regulators, ie the Health and Safety Executive (HSE), the Environment Agency (EA) and the Scottish Environment Protection Agency (SEPA). However, the controls they require are complementary, and the most sensible way of managing both types of risk is through a single management system.

3   This booklet provides guidance on managing the safe segregation, handling, transport and disposal of clinical waste from its point of origin to its point of final disposal. The guidance has been produced jointly by the Health Services Advisory Committee (HSAC) and the Environment Agency, and agreed by the Scottish Environment Protection Agency. It expands and revises the previous Health Services Advisory Committee publication *Safe disposal of clinical waste*, and replaces *Waste management paper No 25 on clinical wastes* (issued by the then Department of the Environment). It takes into account the Carriage of Dangerous Goods (Classification, Packaging and Labelling) and Use of Transportable Pressure Receptacles Regulations 1996 and other recent legislation relating to the carriage of dangerous goods and to waste management.

4   The guidance in this document is intended to help anyone that plays a part in ensuring that risks from clinical waste are minimised. It is aimed at managers responsible for the health and safety of staff and other people. As such, it provides practical advice for both clinical and other staff that are responsible for producing and implementing local waste management policies.

5   This guidance is relevant across the healthcare sector, ie in hospitals, health centres, nursing homes, general practice, ambulance services, dentistry, and in the community. It is also relevant to:

- other producers of clinical waste, for example special schools, research establishments and veterinary practices;
- those in the transport industry who carry the waste; and
- those responsible for the treatment and disposal of the waste.

Who should use this guidance?

6   This guidance provides advice for the following groups of people:

- **Employers and senior managers**, such as chief executives and board members, general practitioners (GPs), nursing home owners, dentists, vets, and directors of companies involved in the transport and disposal of clinical waste. They have the responsibility for setting policies, and for ensuring risks are assessed and eliminated or reduced. It is important that they are seen to be actively committed to ensuring that policies work in practice, and that they support measures required to eliminate or minimise risk. The legal responsibilities of employers are explained in paragraphs 25 to 48.

- **Line managers**, such as ward managers, practice and departmental managers, community health service managers, nursing/residential care home managers/matrons, and transport managers. They are responsible for implementing the clinical waste policy and procedures. They need to know what the policies require from them. This guidance will help them make sure that adequate local management systems are in place, and that there is an upward flow of information to senior managers.

- **Others who can influence policies and practice**, including trainers, occupational health staff, infection control teams and other medical and nursing staff, risk managers, health and safety advisers, estates managers, and safety representatives. Their responsibilities depend on their role. This guidance will be helpful when they are asked to advise on issues relating to clinical waste.

Definition of clinical waste

7   The definition of clinical waste used in this guidance is taken from the Controlled Waste Regulations 1992.[1]

> **Clinical waste means:**
>
> - any waste which consists wholly or partly of:
>   - human or animal tissue;
>   - blood or other body fluids;
>   - excretions;
>   - drugs or other pharmaceutical products;
>   - swabs or dressings;
>   - syringes, needles or other sharp instruments;
>
>   which unless rendered safe may prove hazardous to any person coming into contact with it. And:
>
> - any other waste arising from medical, nursing, dental, veterinary, pharmaceutical or similar practice, investigation, treatment, care, teaching or research, or the collection of blood for transfusion, being waste which may cause infection to any person coming into contact with it.

8   In practice, arrangements for managing clinical waste are based on categorisation of the waste into groups which present different hazards. This guidance is prepared on the basis of five broad groups of clinical waste, which provide a basis for local risk assessments. These are shown in Table 1. These groups do not set out a hierarchy of the relative level of risk from clinical waste; the actual level of risk varies within the groups. The sort of items listed as examples of Group E are usually low risk. However, in some cases they may present a significant risk, on the basis of advice given by an infection control team.

Table 1 Categorisation of clinical waste

Waste group	Type of clinical waste
Group A	Includes the following items: identifiable human tissue,[*] blood, animal carcasses and tissue from veterinary centres, hospitals or laboratories. Soiled surgical dressings, swabs and other similar soiled waste. Other waste materials, for example from infectious disease cases, excluding any in Groups B-E.
Group B	Discarded syringe needles, cartridges, broken glass and any other contaminated disposable sharp instruments or items.
Group C	Microbiological cultures and potentially infected waste from pathology departments and other clinical or research laboratories.
Group D	Drugs or other pharmaceutical products.
Group E	Items used to dispose of urine, faeces and other bodily secretions or excretions which do not fall within Group A. This includes used disposable bed pans or bed pan liners, incontinence pads, stoma bags, and urine containers.[†]

[*] All identifiable human tissue, whether infected or not, may only be disposed of by incineration.
[†] Where the risk assessment shows there is no infection risk, Group E wastes are not clinical waste as defined.

Treated waste

9   Producers of treated waste who do not want to treat it as clinical waste under environmental legislation, need to be able to show the Environment Agency or the Scottish Environment Protection Agency ('the Agencies') that it:

- is safe and non-infectious; and
- cannot be distinguished from other similar non-clinical wastes.

10   In these circumstances, the waste may be classed as non-clinical waste. The most effective kind of evidence is documentary proof that the waste has been suitably and successfully treated to ensure the removal of any hazard to anyone who may come into contact with it.

Special waste

11 Some clinical waste is also classified as 'special waste', and subject to controls under the Special Waste Regulations 1996.[2] These apply over and above other waste management controls. The Agencies have assessed types of clinical waste, and their view is that only the following will be subject to the Special Waste Regulations 1996:

- waste containing Advisory Committee on Dangerous Pathogens (ACDP) Hazard Group 4 biological agents.[3] It is rare that any untreated pathogens in this category are transported. An example of where clinical waste needs to be consigned for disposal as special waste is waste from a viral haemorrhagic fever patient, cared for initially at premises where there are no facilities for autoclaving. This waste must be treated as special waste; and
- waste containing or consisting of prescription-only medicines.[4]

Used sharps and/or fully discharged syringes may still contain or be contaminated with prescription-only medicines and therefore be special waste. The Agencies have issued technical guidance offering further advice on what wastes they consider to be special.[5,6]

Waste from human hygiene

12 Some waste from human hygiene may carry micro-organisms. Examples include:

- sanitary towels;
- tampons;
- nappies;
- stoma bags;
- incontinence pads; and
- other similar wastes, provided that they do not contain sharps.

The following wastes may also contain micro-organisms:

- pregnancy kits;
- blood cholesterol testing devices; and
- condoms.

When such waste is generated in the home, the source population is generally healthy, and the waste is not considered to be either infectious or clinical waste. The householder may put it into the domestic waste, provided it is adequately wrapped and free of excess liquid.

13 Similar wastes may also be generated in the home by people who are undergoing medical treatment. In these cases, the patient's general practitioner may, in consultation with the local Consultant for Communicable Disease Control,[*] identify a specific risk

---

[*] In Scotland, Consultant in Public Health Medicine (Communicable Disease and Environment Health)

associated with the illness, and make appropriate arrangements for disposal of the waste. If there is no specific risk associated with the patient's illness, then such wastes are not clinical waste, since they do not present a significant risk, either to family members or others. The householder may put such waste into the domestic waste, provided it is adequately wrapped and free of excess liquid.

14 Human hygiene waste is also generated in, for example:

- child care centres;
- schools;
- nurseries;
- motorway service areas;
- airports;
- shopping malls;
- offices;
- factories; and
- sports complexes.

Here again, the usual assumption is that such waste is not clinical waste, since the source population is essentially healthy, and the risk of infection is no greater than that from domestic waste. However, those doing the risk assessment may have local knowledge which means they cannot make this assumption.

15 Although human hygiene wastes from these sources may be non-clinical wastes, in quantity they can be offensive and cause handling problems, particularly on collection. In order to assist producers and those collecting waste in meeting the requirements of the duty of care (see paragraph 36), it is considered appropriate to package human hygiene waste separately from other waste streams, where the premises generate more than one standard bag/container of this waste, over the normal collection interval. An example of a commonly used colour coding system is given in paragraph 72. Such wastes may still be collected and disposed of as part of the municipal waste stream.

16 Residential homes[*] and nursing homes[†] cannot assume a healthy source population. Risk assessment will show that their waste falls into two groups:

- those items or categories of items, that represent a significant risk of infection; and
- those that do not.

Those in the first group are clinical waste and should be treated as such, while those in the second group may be treated as domestic waste. However, the offensiveness of substantial quantities of non-infectious waste needs to be taken into account when deciding on how to package the waste (see paragraphs 15 and 72).

---

[*] Residential homes are premises registered under the Registered Homes Act 1984 Part 1 Residential Care Homes and in Scotland, under the Social Work (Scotland) Act 1968 as amended.

[†] Nursing homes (including acute independent hospitals) are premises registered under the terms of the Registered Homes Act 1984 Part 2 Nursing Homes and Mental Nursing Homes, and in Scotland, the Nursing Homes Registration (Scotland) Act 1938.

Human tissue

17 Identifiable human tissue is clinical waste unless it has been rendered safe and non-infectious. After treatment, identifiable human tissue remains an offensive waste. It therefore needs to be dealt with by an authorised waste handling facility. The only appropriate treatment for identifiable human tissue is incineration.

Definition of clinical waste for transport

18 Those involved in the transport of clinical waste need to know that the law concerning the carriage of dangerous goods now covers the transport of infectious wastes. However, the definition of clinical waste for transport is different from the definition used in the Controlled Waste Regulations.[1]

19 The Carriage of Dangerous Goods (Classification, Packaging and Labelling) and Use of Transportable Pressure Receptacles Regulations 1996[7] (CDGCPL2) require (among other things) the proper classification of clinical waste according to its main hazard. For infectious substances, the main hazard for carriage is always its infectious property, even though it may also possess other hazardous properties such as toxicity or flammability. Once the main hazard has been identified, other information is required in order to identify the correct UN number. The UN number is a four digit serial number, devised by the United Nations as a means of identifying dangerous goods, and dictates the method of transportation.

20 The UN number can be identified by referring to two approved documents made under CDGCPL2. These are the:

- *Approved Carriage List: Information approved for the carriage of dangerous goods by road and rail other than explosives and radioactive material* (Approved Carriage List);[8] and
- *Approved requirements and test methods for the classification and packaging of dangerous goods for carriage* (Approved Requirements).[9]

21 There are three UN numbers under which infectious waste might be carried:

- UN 2814 - INFECTIOUS SUBSTANCE, AFFECTING HUMANS;
- UN 2900 - INFECTIOUS SUBSTANCE, AFFECTING ANIMALS; and
- UN 3291 - CLINICAL WASTE UNSPECIFIED, (N.O.S). Not otherwise specified
    REGULATED MEDICAL WASTE, (N.O.S).
    (BIO)MEDICAL WASTE, (N.O.S).

The first two are used where the pathogen with which the waste is contaminated can be specified and is in risk groups 2 - 4 as defined in the Approved Requirements. Most clinical waste is transported under UN 3291 which applies where the nature of the pathogens contaminating the waste cannot be specified; this includes most laboratory waste other than culture plates of known biological agents.

22 These UN numbers should only be used for transporting infectious clinical waste. Different UN numbers apply to the transport of other clinical waste such as waste pharmaceuticals. For example UN 1851 (MEDICINE, LIQUID TOXIC N.O.S.) might be used where the wastes are toxic and liquid.

23 As far as CDGCPL2[7] is concerned, the following wastes described in the Approved Requirements are defined as **not** dangerous for carriage:

- 'sanpro waste', which is defined as including any item of waste used for the collection or disposal of human excreta or secreta, as well as sanitary towels, tampons, nappies and incontinence pads, but excluding waste from the medical treatment of humans. Sanpro waste is also excluded from the definition of clinical waste for carriage;
- healthcare waste, which is defined as any item of waste which arises from the medical treatment of humans or animals, other than healthcare 'risk' waste; and
- decontaminated wastes, which previously contained infectious substances, ie wastes which have undergone decontamination by a process that renders them safe and non-infectious.

24 The following wastes described in the Approved Requirements[9] have been designated as healthcare 'risk' waste and will always be considered dangerous for carriage:

- any infectious biological waste, eg human tissue or blood;
- any related swabs and dressings from hospitals, clinics, surgeries or laboratories;*
- pharmaceuticals which are toxic, flammable, radioactive or have other hazardous properties;
- any infectious waste known or likely to be contaminated with pathogens in risk groups 2, 3, or 4 as defined in the Approved Requirements; and
- sharps, ie any discarded syringe needles, cartridges, broken glass or any other contaminated disposable sharp instruments or items.

* This section also includes waste produced by similar establishments, for example nursing/residential care homes.

Managing the risks to people - the law

25 Under health and safety law, employers who generate clinical waste must ensure that the risks from it are properly controlled. In practice, this involves:

- assessing the risk;
- developing policies;
- putting arrangements into place to manage the risks; and
- monitoring the way these arrangements work.

Although various tasks will be delegated down the management chain to suitably trained and experienced staff, the responsibility for seeing that the policies are developed and effectively implemented remains with the employer. Arrangements for managing clinical waste need to be part of an employer's overall health and safety management system. The HSAC publication *The management of health and safety in the health services*[10] provides further advice on this.

26 The first step in effective management of clinical waste is the proper identification and assessment of risk. Employers must assess the risks to their employees and others under both the Control of Substances Hazardous to Health Regulations 1999[11,12] (COSHH), and the Management of Health and Safety at Work Regulations 1992[13,14] (the Management Regulations).

27 Although only the courts can give an authoritative interpretation of the law, in considering the application of these regulations and guidance to people working under another's direction, the following should be considered:

If people working under the control and direction of others are treated as self-employed for tax and national insurance purposes, they may nevertheless be treated as their employees for health and safety purposes. It may therefore be necessary to take appropriate action to protect them. If any doubt exists about who is responsible for the health and safety of a worker this could be clarified and included in the terms of a contract. However, remember, a legal duty under section 3 of the Health and Safety at Work etc Act (HSW Act)[15] cannot be passed on by means of a contract and there will still be duties towards others under section 3 of HSW Act. If such workers are employed on the basis that they are responsible for their own health and safety, legal advice should be sought before doing so.

The Control of Substances Hazardous to Health Regulations 1999 (COSHH)

28 The COSHH Regulations provide a framework of actions designed to control the risk from a range of hazardous substances, including clinical waste.

COSHH - key points

Employers must, among other things:

- assess the risks to employees and others from clinical waste;
- make arrangements for reviewing the assessment as and when necessary;
- adequately control those risks;
- provide suitable and sufficient information, instruction and training for employees about the risks; and
- provide health surveillance and immunisation, where appropriate.

The Management of Health and Safety at Work Regulations 1992 (as amended) (The Management Regulations)

29 The Management Regulations[13] and their associated Approved Code of Practice[14] provide a framework for managing risks at work, including risks from clinical waste, not covered by more specific requirements such as COSHH.

The Management Regulations - key points

Employers must among other things:

- make a suitable and sufficient assessment of the risks to employees and others. If they have five or more employees, they must record the significant findings of the assessment;
- take particular account in their assessment of risks to new and expectant mothers and their unborn and breast feeding children;[16]
- take particular account in their assessment of risks to young people;
- make arrangements for the effective planning, organisation, control, monitoring and review of any precautions;
- provide health surveillance where appropriate;
- have access to competent health and safety advice;
- provide information for employees; and
- co-operate with other employers who may share the workplace.

Consulting employees

30 Two sets of health and safety regulations cover the need for consultation with employees. The Safety Representatives and Safety Committees Regulations 1977[17] deal with consultation of recognised trade unions through their safety representatives. The Health and Safety (Consultation with Employees) Regulations 1996[18] apply to consultation with employees who are not covered by trade union safety representatives.

31 In summary, employers must consult employees and their representatives about aspects of their health and safety at work, including:

- any change which may substantially affect their health and safety;
- the employer's arrangements for obtaining competent health and safety advice;
- the information provided on reducing and dealing with risks;
- the planning of health and safety training; and
- the health and safety consequences of introducing new technology.

32 In practice, employers have found that policies for ensuring that clinical waste is dealt with safely are only fully effective if they closely involve employees and their representatives.

The Transport of Dangerous Goods (Safety Advisers) Regulations 1999[19]

33 These Regulations (the TDGSA Regulations) implement a European Directive[20] which requires undertakings, whose activities include the transport (and related loading or unloading) of dangerous goods, to appoint vocationally qualified safety advisers. Implementation measures must be in place by 31 December 1999, and employers must therefore have appointed vocationally qualified safety advisers by then. The TDGSA Regulations came into force on 1 March 1999 to provide time for prospective safety advisers to obtain the vocational training certificate which they will need in order to be appointed.

> **The TDGSA Regulations - key points**
>
> Many employers who are involved in the transport of clinical waste by road or rail need to:
> - appoint vocationally qualified safety adviser(s) to advise them on the health, safety and environmental implications of such activities as well as the related risks to property;
> - ensure that those appointed have a vocational training certificate appropriate to the mode of transport being used and to the goods being transported. Such a certificate will be obtained by passing an approved examination; and

- ensure that the appointed safety adviser(s) are given sufficient time and resources to carry out their duties and functions.

The Regulations and an HSE leaflet *Are you involved in the carriage of dangerous goods by road or rail*[21] contain further information on who needs to appoint such safety advisers, what safety advisers are required to do, and how prospective safety advisers can obtain an appropriate vocational training certificate.

Managing the risks to the environment - the law

34 Policies for the management of clinical waste derive from statutory and policy arrangements for waste in general. The Department of the Environment, Transport and the Regions (DETR), the Scottish Office and the National Assembly for Wales are responsible for waste policy. Their policy is based on the concept of waste hierarchy. This encourages approaches which, while protecting or enhancing the environment, either minimise the production of waste or recover the maximum value from it. Further developments in waste policy may be expected to influence the management of clinical waste.

35 Those who produce, transport, treat and dispose of clinical waste all have duties under environmental law. The most important of these are:

- the 'duty of care' in the management of waste;
- the duty to control polluting emissions to air;
- the duty to control discharges to sewer; and
- the obligations of waste managers.

The Agencies also have an obligation to include among the types of environmental harm anything that may be offensive,[22] ie that smells or looks unpleasant. In addition, the waste producer should review the types of waste produced and disposal options, from the point of view of their long-term sustainability.

Duty of care

36 The statutory requirements covering duty of care in waste management are contained in section 34 of the Environmental Protection Act 1990 (EPA)[22] and the Environmental Protection (Duty of Care) Regulations 1991.[23] Further advice is contained in a statutory Code of Practice[24] and a DOE leaflet.[25]

> **Environmental duty of care - key points**
>
> Waste producers must:
> - supply a written description of the waste which includes:
>   - its nature, source and quantity;
>   - sufficient information to enable people who handle the waste further down the chain to discharge their duty of care; and
>   - anything likely to affect the handling or disposal of the waste;
> - satisfy themselves that the means of treatment and disposal are appropriate to the waste.

37 A key element of the duty of care is keeping track of the waste. The holder of the waste is responsible for:

- taking adequate steps to ensure that the waste is managed safely, and kept secure; and
- transferring it only to an authorised or exempt person.

At different points in the chain, the holder may be producing, carrying, treating or disposing of the waste.

38 When waste is transferred from one party to another, the person handing it on (the 'transferor') must complete a transfer note. The transferor and the recipient (the 'transferee') sign the note; both of them take and keep a copy of it. An 'annual transfer note' may be used to cover all the movements of regular consignments of the same waste, between the same parties.

39 In general, where X carries waste for Y, X must register with the Agencies. Most carriers of controlled waste must be registered.[26] Detailed requirements are imposed by Regulations.[27] An exemption from registration covers:

- people such as community nurses and others working in home healthcare; and
- healthcare providers' vehicles carrying waste generated by them.

40 Householders who carry only household waste, generated by them and transported in their own vehicle, are exempt from the duty of care.

Other legal requirements

41 The Environmental Protection Act 1990[22] (EPA) and the Waste Management Licensing Regulations 1994[28] provide a legislative system to regulate waste management. Waste must be managed in a way which does not cause pollution or harm to human health. Under the legislation, controlled waste (which includes clinical waste) must be kept, treated or disposed of in accordance with a waste management licence, unless it:

- is consigned to a process which is subject to other control regimes in accordance with regulation 16 of the 1994 Regulations; or
- falls within an exemption under Schedule 3 of the 1994 Regulations.

These exemptions mean, among other things, temporary stores (other than waste transfer stations) and laboratory autoclaves do not require a waste management licence.

42 The licences and licence exemptions for individual waste management facilities specify what types of waste they can handle, and set out terms and conditions to prevent pollution or harm to the environment. Incinerators that handle clinical waste are governed by an authorisation under Part 1 of EPA. Depending on the size of the facility, the authorisation may be for either:

- a Part A process covering all emissions (Integrated Pollution Control (IPC)); or
- a Part B process covering emissions only to air (Local Authority Air Pollution Control (LAAPC) or in Scotland (LAPC)).

IPC authorisations are granted by the Agencies. Local authorities grant LAAPC authorisations, while SEPA grants LAPC authorisations in Scotland.

43 EPA[22] also covers offensive waste. Duties in respect of this are to either prevent the offence or to render the waste inoffensive. Most clinical waste would be considered offensive. In the context of healthcare, however, the treatment methods for hazards usually deal with the offensiveness.

44 There is a statutory requirement under the Water Industry Act 1991[29] and the Sewerage (Scotland) Act 1968[30] to control discharges to sewers.

Special waste

45 As mentioned in paragraph 11, a few clinical wastes are subject to more stringent controls under the Special Waste Regulations 1996.[2] Under these Regulations, all movements must be tracked using consignment notes until they reach an appropriate waste management facility.

> **Special waste - key points**
>
> Those responsible for consigning the waste must among other things:
> - consider whether it is necessary to notify the Agencies in advance of their intention to move the waste;
> - complete the relevant sections of the consignment note to accompany the waste being moved;
> - pay the appropriate fee to the Agencies each time the waste is moved, if applicable; and
> - keep adequate records for three years.

46 For standard movements of special waste, the Agency in the area where the waste is going must receive at least three days notice that the special waste is being moved. A fee is payable for most special waste movements.

47 Advance notification is not required for repeated consignments involving the same types of waste between the same consignor and consignee. In such circumstances, one notification can cover a period of 12 months. However, a fresh set of consignment notes is necessary every time special waste is moved.

48 The Regulations allow for a number of collections by the same carrier from different consignors over a 24 hour period (referred to in the legislation as a 'carrier's round'), for which a single notification and fee are required. In addition, a single fee is payable for a succession of carrier's rounds undertaken over a period of one week, subject to a number of restrictions designed to limit the provision to businesses carrying out small-scale operations. In these circumstances, it is the carrier who completes the consignment notes. However, the producer must still keep adequate records. During 1999, DETR has been reviewing the Regulations. It has been considering a proposal to extend the time allowed for a carrier's round to 72 hours.

Precautions for dealing with clinical waste

49 The precautions required when handling clinical waste depend on the results of risk assessment, and the relevant legal requirements. The following paragraphs provide guidance on the range of measures that need to be considered in relation to:

- training and information; ✓ SAC. Procudure (check 01/02)
- personal hygiene; ✓ Procadure
- personal protective equipment; ✓ "   "
- immunisation;
- segregation; ✓ "   "
- handling; ✓ "   "
- packaging;
- labelling; ✓ PROCEDURE
- storage;
- transport on site and off site;
- accidents, incidents and spillages; and ✓ PROCEDURE
- treatment and disposal.

50 Such precautions need to operate within an overall policy for risk management, which covers clinical waste. To be successful, the policy needs to be actively supported by detailed procedures for clinical waste. Line managers and others need to have clearly defined responsibilities, and someone, such a waste manager, needs to properly monitor the arrangements.　　　　　　　　　　　　　　　　　　　　　　　　　PROCEDURE P.2,

51 In hospitals, the following staff are likely to be among those who need to be consulted when developing policy and arrangements for handling and disposal of clinical waste:

- employee representatives;
- infection control doctor;
- infection control nurse;
- senior nursing personnel;
- heads of domestic and portering services;
- health and safety officer;

- head of pharmaceutical services;
- occupational health staff;
- contractors providing relevant services;
- director/head of estates;
- director/head of facilities;
- officer designated for overseeing waste management on site, for example the waste manager;
- relevant line managers; and
- risk management team.

Smaller establishments generating clinical waste may not have this range of expertise available to them, but should still have access to competent advice on clinical waste issues.

52 Local procedures need to:

- be written in a way which can be understood by those who need to follow them, including those who may not have a good command of English;
- take account of different levels of training, knowledge and experience;
- be up to date;
- be available to all staff including part-time, shift, temporary, agency and contract staff;
- be available in all areas.

53 Managers need to ensure that procedures are followed by all staff. Staff at all levels who generate the waste need to recognise that they are personally responsible for complying with agreed local procedures.

Training and information

54 The risk assessments required by the Management Regulations[13] and COSHH[11] should identify which staff are involved in handling clinical waste. Under the Health and Safety at Work etc Act 1974,[15] the Management Regulations and COSHH, they must receive information on:

- the risks to their health and safety;
- any precautions necessary;
- the results of any monitoring carried out; and
- the collective results of any relevant health surveillance.

A training record will readily enable line managers to identify members of staff who are not receiving the appropriate level of training, and where to focus such training.

55 It is important to provide training for all staff identified as at risk, not just obvious groups such as medical and nursing staff. Staff who change location may also require appropriate training/retraining.

56 Training needs vary depending on the job and on the individual. All staff involved in handling clinical waste need training, information and instruction in:

- the risks associated with clinical waste, its segregation, handling, storage and collection;
- personal hygiene;
- any procedures which apply to their particular type of work;
- procedures for dealing with spillages and accidents;
- emergency procedures; and
- the appropriate use of protective clothing.

57 For staff who collect, transfer, transport or handle quantities of clinical waste, the training needs to cover:

- checking that storage containers are sealed effectively before handling;
- ensuring that the origin of the waste is marked on the container;
- handling sacks correctly (for example, not clasped to the body, thrown, dropped or supported by hand from below);
- using handles to move rigid containers;
- checking that the seal on any used waste storage container is unbroken when movement is complete;
- special problems relating to sharps disposal;
- procedures in case of accidental spillage and how to report an incident; and
- safe and appropriate cleaning and disinfection procedures.

58 Some staff require more specific training; these include incinerator operators, drivers, community and laboratory staff. Under the Environmental Protection Act 1990 section 74,[22] some incinerator operators and all landfill and waste treatment operators require a certificate of competence from the Waste Management Industry Training and Advisory Board. Drivers of vehicles that transport clinical waste by road may need additional training under the Carriage of Dangerous Goods by Road (Driver Training) Regulations 1996 (DTR2).[31] Prospective safety advisers who are appointed under TDGSA Regulations[19] may need to undertake training before sitting the approved examination (see paragraph 33).

59 People using protective equipment must receive training on: the risks which the equipment is designed to avoid or limit; the way in which it will be used; and the procedures for ensuring maintenance and repair.

60 Trainers need to be competent and be able to communicate their expertise. Occupational health advisers, infection control staff and safety representatives have a key role in providing information and developing suitable training materials.

Personal hygiene

61 Basic personal hygiene is important in reducing the risk from handling clinical waste. Employers need to ensure that washing facilities are convenient for people handling clinical waste; this is particularly important at storage and incineration facilities.

Personal protection

62 Employers must not use personal protection as the first line of defence against health risks, unless all other reasonably practical precautions have been taken. Risk assessments should identify situations in which the hazard cannot be adequately controlled by any other means, and personal protective equipment is required. In such cases, employers must ensure that these items are provided, used and maintained. They must also make appropriate arrangements for storage and cleaning. Under the law, employees must co-operate with employers to ensure that their legal duties are met.[32]

63 A risk assessment of the work done by staff who regularly handle, transfer, transport, treat or incinerate clinical waste is likely to show that they require:

- suitable heavy-duty gloves when handling clinical waste containers;
- safety shoes or industrial wellington boots to protect the feet against the risk of containers being accidentally dropped. The soles of such shoes or boots may also need to provide protection against the spillage of sharps and slippery floors; and
- an industrial apron or leg protectors if container handling creates a risk of bodily contact.

64 Staff require protective face visors, helmets and strong industrial gloves where incinerators or other machines are charged manually. They should wear suitable respiratory protection to an appropriate standard for protection against toxic dusts during the removal of ash and/or residue from the incinerator. COSHH[11] assessments will identify any other circumstances during which respiratory protection is required.

65 For situations such as cleaning spillages, the risk assessment may indicate a need for protective equipment to prevent skin contact. In these cases, disposable gloves and aprons are best. In some circumstances, face visors may be necessary to protect employees from splashing. Such instances should be identified in local procedures.

66 Surgical masks do not provide the protection against micro-organisms required under COSHH.

67 Employers need to take account of possible risks from latex allergy when selecting gloves.[34,35]

Immunisation

68 Staff handling clinical waste should be offered appropriate immunisation, including hepatitis B and tetanus.[36] Employers need to establish arrangements for dealing with staff who decline the immunisation services that are offered and those who do not sero-convert. These arrangements should include:

- advice about the risks; and
- clear guidelines for the staff concerned, for example explaining that this may prevent them from carrying out certain work, such as conducting exposure prone procedures on patients.

Segregation

69 Proper segregation of different types of waste is critical to safe management and helps control disposal costs.[37] Wastes need to be sorted at the point of origin so they can follow appropriate routes for treatment and/or disposal. Segregation only works if staff are provided with:

- background information and reasons for segregation;
- appropriate equipment, such as sufficient colour-coded waste containers;
- clear instructions and training.

It is essential that the procedures used for segregating waste are monitored and evaluated on a regular basis, and that the staff involved receive feedback on how the arrangements are working.

70 Everyone working in areas where clinical waste arises needs to receive clear information, instruction and training on categorising waste. It is helpful if posters showing the different waste streams and types of waste are displayed at appropriate locations.

71 Implementing a system for segregation of clinical waste streams may involve significant changes in waste management practices. Preparation is essential to:

Safe disposal of clinical waste

- ensure that staff are involved in the process of change;
- ensure that non-clinical wastes are redirected from the clinical waste stream; and
- minimise the risk of system failure.

72 A system of colour-coding aids the process of waste segregation. The following system for colour-coding is widely used for waste containers:

- **yellow** - Group A clinical wastes which will be incinerated in a clinical waste incinerator, or otherwise disposed of. If the producer needs to distinguish between disposal methods, another colour may be used for this waste. For example, in Scotland, orange bags are used for dressings, incontinence pads, dialysis waste, blood and liquid clinical waste, and sterilised high-risk laboratory waste which is sent for heat treatment before disposal;

- **yellow with black stripes** - non-infectious waste, eg Group E and sanpro which is suitable for landfill or other means of disposal;

- **light blue or transparent with blue lettering** - waste for autoclaving or equivalent treatment before disposal; and

- **black** - treated clinical waste, non-clinical or household waste.

In addition to colour-coding, the containers need to be of the correct standard for the waste they will contain (see paragraphs 94-108).

73 Aerosols and other pressurised containers should not be placed in clinical waste containers destined for incineration, but in separately identified containers specifically for them. Aerosols may pose a safety and/or environmental risk for a number of reasons, for example:

- they may contain:
  - CFCs;
  - prescription only medicines; or
  - flammable liquids;
- explosions from aerosols may damage the refractory lining of incinerators.

The local waste management procedures need to clearly set out the arrangements for disposal of these items. Used or expired pharmaceutical aerosols are special waste and need to be returned to the pharmacy for disposal.

[Handwritten margin note: SCOTLAND ORANGE. J NICHOLSON - OK]

Handling clinical waste before disposal

74 While segregation provides the basis for the safe handling of wastes, other precautions are needed to ensure that risks are managed as the waste moves downstream from source to final disposal.

Group A wastes

75 Risk assessments are likely to indicate that the procedures for Group A wastes should include the following measures:

- placing waste in waste sacks in sack holders or other appropriate containers at the point of generation;
- replacing sacks daily or when three-quarters full;
- not transferring loose contents from container to container;
- sealing sacks with a plastic tie, closure or heat sealers, purpose-made for clinical waste sacks;
- prohibiting the use of staples to close the sack, as they do not provide a secure closure and may puncture the sack;
- providing lids for bins which can be sealed before collection;
- labelling sacks to indicate their origin, for example by coding on the sack itself, by suitable permanent marker, by a label showing clearly the name of the hospital and the department, or by pre-printed self-adhesive labels or tape; and
- collecting at appropriate frequency.

76 Procedures need to be particularly robust where large quantities of waste and identifiable human tissue are generated, for example in theatres.

77 Where waste accumulates in small quantities daily, the interval between collections ought to be as short as reasonably practicable and preferably not less than once a week. As elsewhere, containers need to be properly sealed, labelled and kept secure before removal.

Group B wastes (including broken glass)

78 Many needlestick injuries happen during resheathing. Risk assessment shows that needles should not be resheathed by hand. Syringes/cartridges and needles should be disposed of intact. Sharps must never be placed in containers used for the storage of other wastes. They need to be put safely into properly constructed sharps containers. Such containers should either meet the requirements of BS 7320[38] and/or be UN type-approved (see paragraphs 100-104).

79 While in use, sharps containers should be kept out of the reach of members of the public, in particular small children and people who may not appreciate the risks associated with this type of waste. To avoid damage by heat, sharps boxes should not be placed near radiators or in direct sunlight.

80 Experience also indicates that particular attention needs to be paid to the provision of sufficient sharps boxes in a number of areas, for example theatres, accident and emergency and out-patients departments, to ensure that they contain all the disposable sharps generated.

81 To avoid the risks associated with overfilling, sharps containers need to be removed when three-quarters full, sealed, and labelled. They should not be placed in sacks; keeping them separate during storage and transport ensures that any faulty or broken containers, which may leak fluid or allow penetration of sharps, are more obvious. Any damaged containers need to be placed in a larger secure, rigid container which is properly labelled.

82 Used needles and syringes must not be disposed of in domestic waste. Healthcare staff who treat patients at home should place any sharps and syringes that they generate in appropriate containers for disposal, either through their employer's clinical waste disposal system, or for collection as appropriate.

83 A variety of devices are available which destroy needles, reducing the risk of injury. Employers may wish to consider using such devices, if their risk assessment identifies a particular need. Some types of device are widely used by people who regularly self-inject at home. They should not be disposed of in the domestic waste. Disposal arrangements for these and used syringes can be made with local hospitals, health centres, clinics, local authorities or pharmacists.

Group C wastes

84 Risk assessment is likely to indicate that waste from research or clinical laboratories may contain pathogens. Waste known or likely to contain ACDP Hazard Group 3 and 4 biological agents[3] and cultures of ACDP Hazard Group 2 biological agents should be made safe, by autoclaving, before leaving the premises for final disposal.* In exceptional circumstances, for example an autoclave malfunction, waste which is normally autoclaved, should be packaged in accordance with the *Approved Requirements* for Carriage,[9] and transferred to an incinerator as soon as possible. It should not be allowed to accumulate for more than 24 hours.

* Any waste from TSE agents should be disposed of in accordance with current guidelines.[39]

Group D wastes

85 These wastes are likely to be special waste and should normally be returned to a responsible person at a hospital or community pharmacy, who should make arrangements for their safe disposal, for example by using a specialist contractor.

86 The disposal of controlled drugs is regulated by the Misuse of Drugs Regulations 1973[40] and the NHS Executive has produced guidance on the topic.[41]

Group E wastes

87 The contents of disposable items in this category, such as excreta, may be discharged to the sewer, via the sluice, WC or purpose-built disposal unit. These items do not normally fall within the definition of infectious waste for transport purposes and therefore do not have to be packaged in UN type-approved containers.

88 Households which generate large amounts of Group E waste may obtain advice on its collection from the local authority.

Storage

89 Clinical waste containers may need to be stored before incineration or transport for disposal elsewhere. They should not be allowed to accumulate in corridors, wards or other places accessible to members of the public. Large establishments may need small satellite stores.

90 The risk assessment for a bulk clinical waste storage area is likely to indicate that it should be:

- reserved for clinical waste only;
- well lit and ventilated;
- close to any on-site incineration or other disposal facility;
- sited away from food preparation and general storage areas, and from routes used by the public;
- totally enclosed and secure;
- provided with separate storage for sharps containers, which may need a higher degree of security to prevent unauthorised access;
- sited on a well-drained, impervious hardstanding;
- readily accessible but only to authorised people;
- kept locked when not in use;

Safe disposal of clinical waste

- secure from entry by animals and free from insect or rodent infestations;
- provided with wash-down facilities;
- provided with washing facilities for employees;
- clearly marked with warning signs;
- provided with separate clearly labelled areas for containers going to landfill, treatment and incineration; and
- provided with access to first-aid facilities.

91 All bulk stores should have storage capacity to match the proposed frequency of collection. Bank or other holidays need to be taken into account, and a margin provided for any interruption in the disposal system.

92 Anybody shut inside storage areas or refrigerators with self-closing doors should be able to open them from the inside.

93 Appropriate protective equipment, including gloves, overalls and materials for dealing with spillages, should be readily available close to storage sites.

Packaging

94 Requirements for packaging vary depending on, for example, the category of the waste, whether it is transported off-site, and the method of final disposal. This section outlines some of the factors that producers need to consider when making decisions about packaging. Appendix 1 provides a summary of packaging requirements.

Packaging of Group A waste - clinical waste sacks

95 Most infectious waste (other than contaminated sharps) is disposed of in clinical waste sacks. If waste containing Group A material is transported off-site for disposal, the sacks need to comply with the requirements of CDGCPL2[7] and the Approved Requirements.[9] This means that they need to be design type-tested and certified (usually referred to as 'UN type-approved'). Treated and/or sterilised clinical waste is not considered dangerous for carriage, and the requirements do not apply to such waste.

96 The provision for the use of type-approved sacks for clinical waste is a relaxation from the terms of the relevant European Union Directive. This relaxation expires on 31 December 2001; United Kingdom law will need to require the packaging of UN 3291 clinical waste in suitable approved rigid packaging, for transport off-site after that date. This rigid packaging is combination packaging, which means that both the rigid outer and any inner plastic packagings have to be type-approved together.

97 Before 1 January 2002, the following options are available when transporting waste classified as UN 3291:

- to continue transporting the waste in type-approved clinical waste sacks only;
- to transport the waste in type-approved clinical waste sacks placed in non-type approved rigid containers; or
- to transport the waste directly in suitable type-approved rigid containers as they become available.

98 Duty holders will probably find it easier to phase in the changes in packaging before the requirements come into force in 2002, rather than try to introduce them all at once. The risks of infection and needlestick injury from the handling of waste sacks are considerably reduced when the sacks are placed in wheely bins or similar containers. This is particularly true if the bins can be handled mechanically, for example when loading vehicles and disposal facilities. The use of sacks to contain infectious clinical waste will still be allowable after 1 January 2002, provided that they are suitable for use in the type-approved rigid packaging (see paragraph 96), or they are not used to transport the waste on a public road.

99 If waste is not transported on public roads, the requirements for type-approval of packaging do not apply. However the risks associated with handling sacks are minimised when the sack is placed in a rigid container, and to comply with COSHH,[9] waste producers need to ensure that clinical waste is packaged safely. To prevent any confusion, it is good practice to use UN type-approved containers, whether or not the waste is being transported off-site.

Packaging of Group B waste - sharps containers

100 If Group B waste is taken off-site for disposal, the sharps containers need to conform to the standards contained in the Carriage of Dangerous Goods Regulations. The majority of sharps waste is transported under UN 3291. Packagings for such waste need to be type-approved in accordance with CDGCPL2[7] and meet the Approved Requirements[9] for the packaging of sharps contaminated with unspecified infectious substances. Where the sharps are contaminated with infectious substances which can be specified and are being transported under UN 2814 or 2900, the packaging requirements are different and are detailed in the Approved Requirements.

101 Some sharps waste may be hazardous in other ways. For example, it may be contaminated with cytotoxic drugs or other potentially harmful pharmaceuticals. The use of sharps boxes, UN type-approved for UN 3291 waste, will still give adequate protection in these circumstances. It should be remembered that such waste is also likely to be *special waste*.

102 If sharps waste is not transported on public roads, producers do not need to use type-approved containers. However, containers need to be safe, and as a minimum need to comply with British Standard BS 7320[38] or equivalent.

103 If there is any risk of confusion between waste for off-site or for on-site disposal, it is good practice and probably simpler to use only containers which are both UN type-approved, and meet BS 7320. Sharps containers which meet both the standards for type-approval under CDGCPL2 and BS 7320 (or equivalent) are readily obtainable.

104 Waste producers do not need to use type-approved containers for waste which is treated and/or sterilised before transport.

Packaging of Group C wastes

105 In most cases, Group C wastes are sterilised on-site before final disposal. Such waste is not considered dangerous for carriage, and there are no special packaging requirements under CDGCPL2.[7] Packaging, nevertheless, needs to be suitable for containing the waste, and the waste must be clearly marked as having undergone sterilisation before leaving the site.

106 Any Group C waste which is not sterilised before leaving the site has to be packaged in accordance with the Approved Requirements.[9]

Packaging of Group D wastes

107 Group D waste for off-site disposal which has been classified as dangerous for carriage, must be transported in appropriate UN type-approved packaging. The use of clinical waste sacks is not acceptable for the disposal of dangerous Group D waste, such as cytotoxic drugs.

108 UN type-approved packaging is not required for Group D waste which is not taken off-site. However, the packaging needs to be suitable for the product involved, and there should be no risk of accidental contact with hazardous substances. As with sharps containers, producers may find it easier to use UN type-approved containers to prevent any confusion between waste for on-site and off-site disposal.

Packaging of Group E wastes

109 Group E wastes are not considered dangerous for transport by road under CDGCPL2 and do not have to be packaged in UN type-approved containers. Risk assessments under COSHH[11] may, however, indicate the need for precautions to ensure the health and safety of people handling the waste. Producers need to take this, and the possibility of causing offence, into account when making decisions about packaging.

Labelling

110 Packaging for clinical waste needs to be labelled to show clearly the type of waste, the hazards and, if appropriate, the method of disposal. If the waste is transported off-site for final disposal, labels must meet the requirements of CDGCPL2. These Regulations prohibit consigning the waste for transport unless the following information is clearly shown:

- the appropriate designation of the waste. For example, the designation of unspecified infectious waste is 'CLINICAL WASTE, UNSPECIFIED, N.O.S', '(BIO)MEDICAL WASTE, N.O.S.' or 'REGULATED MEDICAL WASTE, N.O.S.';
- the appropriate UN number. In the case of 'CLINICAL WASTE, UNSPECIFIED, N.O.S.' this would be UN 3291. The number must always be preceded by the letters 'UN'; and
- the appropriate danger sign. The danger sign for infectious substances which is needed for waste classified as UN 3291 is shown below.

111 CDGCPL2[7] also sets out methods of labelling. It requires the specified information to:

- be positioned so that it can be read easily when the package is loaded for transport;
- stand out from the background;
- be clearly and indelibly marked either directly on the package, or on a label securely fixed to the package.

112 Packaging should be marked in accordance with the Approved Carriage List[8] to indicate any subsidiary hazard. For example, waste medicines have a subsidiary toxic hazard as well as the primary hazard of being flammable liquid. Thus the packaging for UN 3248 must show both the danger sign for flammable liquids and the subsidiary hazard sign for toxic substances. There are exemptions from this requirement when limited quantities are carried. The limits are specified in Schedule 3 of CDGCPL2.

113 Danger signs must conform in form, colour and size to the requirements of CDGCPL2. For example, the danger sign for infectious substances should be black on

white. A relaxation is allowed in the case of the yellow plastic sacks commonly used for the transport of clinical waste; in this case, the danger signs may be printed directly on to the sacks (ie black on yellow). The sides of a danger sign need to be at least 100 mm long. This applies unless the package is an awkward shape, or too small. In such cases, the signs need to be as large as practicable.

114 The requirements for transport off-site do not apply to waste which remains on-site. However, in such cases the packaging still needs to indicate the hazard and the method of final disposal.

Transport

Internal transport

115 Dedicated trucks, trolleys or wheeled containers are needed to transport waste containers to storage areas. In order to prevent contamination they should not be used for any other purpose. They need to be designed and constructed so that they:

- are easy to clean and drain;
- contain any leakage from damaged receptacles or containers;
- are easy to load and unload;
- do not offer harbourage for insects or vermin; and
- do not allow particles of waste to become trapped on edges or crevices.

116 Containers for on-site transport need to be steam cleaned or disinfected following leakages or spills, and at regular intervals. If containers are heavily used, cleaning is likely to be required at least weekly. The clinical waste procedures need to specify the method and frequency of steam cleaning or disinfection.

117 Some commercial companies and hospitals place filled and sealed clinical waste bags and sharps boxes into containers for transport close to source. These containers take waste to disposal facilities where they are automatically discharged. The containers used include wheeled, rigid, lidded plastic or galvanised bins. This has the advantage of reducing the handling of filled containers. Such systems currently represent good practice. **The use of rigid containers for infectious waste will become a legal requirement for transport off-site on 1 January 2002 (see paragraphs 94-99).**

Transport off-site

118 While CDGCPL2[7] covers the classification, packaging and labelling of clinical wastes, a number of other requirements also apply to transport of wastes off-site. The Carriage of Dangerous Goods by Road Regulations 1996[42] (CDGRoad) place duties on those

involved in the carriage of dangerous goods. Most of the duties fall on the consignor, the operator of the vehicle and the driver. People who design and manufacture vehicles intended for the carriage of dangerous goods also have duties. Goods such as clinical waste assigned to UN 3291 are Packing Group II goods under CDGRoad. This means that most of the requirements in CDGRoad apply only if the capacity of each container being transported is greater than 10 kg, and the total vehicle load exceeds 200 kg. A few of the requirements apply at different thresholds.

Transport of packaged clinical waste - key points

The consignor:
- must ensure that the operator receives information about the load to be transported, including designation, classification, UN number and packing group, as well as the number and capacities of the packages and the name and address of the consignor. For goods in Packing Group II, for example UN 3291 Clinical waste, N.O.S., this requirement applies only if goods are in receptacles over 10 litres/10 kg.

The vehicle operator must ensure that:
- any vehicle used for the transport of clinical waste is suitable for the purpose and is adequately maintained. In addition, where clinical waste is carried in road tankers or by vehicles carrying tank containers, the vehicles should meet the appropriate requirements specified in the Approved Vehicle Requirements;[43]
- the driver receives specified information. As well as the information provided by the consignor, this must include details of the complete load. The operator is required to keep this information for at least three months after the relevant journey;
- the driver is trained in accordance with the requirements of Carriage of Dangerous Goods by Road (Driver Training) Regulations 1996[31] (DTR2);
- the goods are loaded, stowed and unloaded in a way that does not create a risk or significantly increase any existing risk to the health;
- the vehicle is marked at the front and rear with an orange plate; and
- adequate precautions against fire or explosion are in place, this includes the provision of fire extinguishers.

The driver must ensure that:
- when the vehicle is not being driven it is parked safely and properly supervised. For Packing Group II goods this only applies when the total load size exceeds 2000 litres/2000 kg; and
- information about the goods being carried is produced when required by an HSE inspector, the police or any goods vehicle examiner.

119 Other HSE guidance provides more detailed information about requirements relating to the transport of dangerous goods.[44,45]

Dealing with accidents, incidents, and spillages

120 Employers at all points in the waste chain need written procedures for dealing with accidents or incidents which cover:

- immediate first-aid measures;
- immediate reporting to a responsible designated person;
- recording of the accident/incident;
- investigating the incident, and implementing remedial action. Initial investigation should preferably take place before any damaged container is removed;
- retaining, if possible, the item and information about its source to help identify possible infection risks;
- attending by any injured person at an accident and emergency department or occupational health department as soon as possible;
- involving the risk manager;
- involving the waste manager; and
- involving the infection control team.

121 In the case of sharps injuries, procedures also need to cover arrangements for suitable medical advice and counselling. Appendix 2 contains an example of information for staff on first aid for sharps injuries.

122 Detailed advice on first aid is available in other HSE publications[46] and DOH guidance.[47]

123 Employers need clear written procedures for dealing with spillages which specify the following:

- reporting and investigation procedures;
- use of a safe system of work for clearing up the clinical waste;
- appropriate requirements for decontamination; and
- protective clothing.

124 The ready availability of appropriate spillage kits helps ensure the correct action in the event of a spillage. Such kits are particularly useful at storage and incineration sites. Spillage kits may contain, for example:

- disposable gloves;
- a disposable apron;

- a clinical waste sack;
- paper towels;
- disposable cloths;
- disinfectant recommended, for example, by the local control of infection policy; and
- a means of collecting sharps.

125 Vehicle operators are required to ensure that vehicles are equipped so that the driver can take any necessary action which is required by the emergency information relating to the load. This may involve having a spillage kit available.

126 Disinfectants containing 10 000 ppm available chlorine are recommended for spillages.[48] The use of sodium dichloroisocyanurate (NaDCC) granules is also generally recommended, because made-up solutions lose activity with time and require regular replacement. Suitable inert, absorbent materials may be used to deal with liquid spillages after disinfection. The infection control team can advise on disinfectants. They should be consulted after a spillage containing or suspected to contain unusual infective agents, for example new variant CJD for which the above disinfectants are ineffective.

127 The use of disinfectants themselves may present a health risk, particularly in confined spaces. COSHH[11] applies to such risks and appropriate precautions must be specified, based on risk assessment. Glutaraldehyde must not be used.

128 Employers need to provide appropriate equipment for collecting spilled waste and placing it in new containers. Sharps must not be picked up by hand. Spilled waste and any absorbent materials need placing in a clinical waste container for disposal. The incinerator operator should be informed before large quantities of non-combustible absorbent material, used for clearing a spillage, are sent.

Treatment and disposal

129 Clinical waste may be treated in a number of different ways before disposal. Treatment reduces the risk of infection during handling and transport, and may also be used to alter the state of the waste so that it is no longer clinical waste by definition. This increases the range of final disposal options. Once adequately treated, what was formerly clinical waste can be sent direct to landfill or incinerated in a municipal waste incinerator.

Treatment methods

130 There are a number of thermal and chemical systems available to treat clinical waste. Some systems are suitable for use on the site where the waste is generated. The selection of the most appropriate system depends, among other things, on the composition and volume of the waste, and on the costs of setting up and operating the system. Further advice on clinical waste technologies can be found in HTM 2075 *Clinical waste disposal/treatment technologies*.[49]

Standards and procedures

131 Any treatment method needs to be reliable and consistently achieve the claimed standard of treatment. Its performance needs to be measurable, and the process controlled precisely enough to reproduce the target standard. Treatment techniques also need to be compatible with the means of final disposal, for example autoclaving microbiological material before landfilling it.

132 Managers of waste treatment plants need to work to audited procedures. This is particularly important if the treatment is intended to reduce the number of, or eliminate, infectious organisms. Procedures need to take into account the risks to operators, as well as the need to maintain standards of waste treatment. The Agencies are establishing standardised testing methods for non-incineration treatment techniques.

133 Waste producers placing clinical waste with contractors for treatment need to consider health and safety aspects before doing so, and have arrangements to monitor these as part of the review of the contractors' performance.

134 With a few exceptions, waste treatment processes need a waste management licence or other authorisation. Where there is any doubt, the operator should consult the Agencies.

Heat treatment

135 Thermal treatment processes maintain the temperature of the waste at specified values for specified lengths of time to reduce the numbers of infectious agents in the waste. Typically, waste is macerated and moistened before being heated. The operating parameters needed to ensure satisfactory inactivation of infectious agents are likely to vary from plant to plant. Heat inactivation systems can be divided into those using relatively low temperatures (such as autoclaves and microwaves), and those using high temperatures (such as pyrolysis, plasma technology and gasification).

Chemical treatment

136 Chemical treatment methods include disinfection with sodium hypochlorite, chlorine dioxide, peracetic acid and calcium oxide. It is important that the waste is shredded beforehand, to bring all surfaces into contact with the chemicals.

Disposal methods

137 There are three principal disposal methods: discharge to sewer; incineration; and landfill.

Discharge to sewer

138 The sewerage system is designed and operated to accept infectious material in the form of domestic sewage. Any discharge to sewer, other than domestic sewage, must have the prior agreement of the statutory responsible body. Table 2 shows the present allocation of the statutory responsibilities for controlling discharges to sewer. Anybody intending to dispose to sewer any waste which may present a substantially greater risk than domestic sewage should first seek advice from, for example, the infection control team, about whether the waste needs to be made safe. Such potentially high-risk waste includes, for example, waste from pathology and other laboratory departments. An unauthorised discharge to sewer is usually not only an offence, but it can be dangerous to workers in sewers, and have unacceptable affects on the efficiency of the sewage treatment works. Anybody considering making discharges to sewers needs to liaise closely with the sewerage undertaker.

Table 2 Statutory responsibility for controlling discharges to sewer

	Discharges to sewer from	
	prescribed processes[22]	processes other than prescribed processes
England and Wales	Environment Agency and sewerage undertaker	Sewerage undertaker
Scotland	SEPA (in consultation with water authorities)	Sewerage/water authorities

139 Some sanitary products contain small amounts of plastic. Attempting to dispose of them to sewer may cause unacceptable operating difficulties for the sewerage undertaker including:

- blockages in the sewerage system;
- binding of pumps and other equipment; and
- a serious litter problem, not only within the boundary of the sewage treatment works but also the surrounding area, including nearby watercourses.

Pharmaceutical waste

140 Waste pharmaceuticals are likely to be prescription-only medicines and therefore subject to the Special Waste Regulations 1996.[2] They should not be discharged to public sewers.

141 Pharmacies, doctors' surgeries, other healthcare establishments and clinics may hold small quantities of controlled drugs. The destruction of these drugs is controlled by the Misuse of Drugs Regulations 1973.[40,41]

Mercury

142 The disposal of mercury is subject to specific control. Depending on the circumstances, the Trade Effluent (Prescribed Processes and Substances) Regulations 1989,[50] the Water Industry Act 1991[29] and the Water Resources Act 1991[51] might apply to the discharge and release of mercury to the environment. Mercury should not normally be discharged to the sewer. Any discharge to the sewer requires consent from the sewerage undertaker.*

* In Scotland, the relevant legislation includes the Sewerage (Scotland) Act 1968 and part II of the Control of Pollution Act 1964.

143 Dental amalgam is generally recovered by separators or sieves, and disposed of separately or recycled. However, a small amount of amalgam may pass into the drains and reach sewage treatment works. The Department of the Environment, Transport and the Regions (DETR) has issued guidance to sewerage undertakers on the control of mercury from hospitals and dentists.[52]

144 Employers who use mercury should carry out a risk assessment for dealing with mercury spillages, and produce written procedures. A spillage kit needs to be available, which includes disposable plastic gloves, paper towels, a bulb aspirator for the collection of large drops of mercury, a vapour mask, a suitable container fitted with a seal and a mercury absorbent paste (equal parts of calcium hydroxide, flowers of sulphur and water). In no circumstances should a vacuum cleaner or aspiration unit be used, as this will vent mercury vapour into the atmosphere. An example of a local mercury spillage procedure is contained in Appendix 4. Mercury amalgam should not be disposed of in sharps containers.

Radioactive waste

145 Radioactive waste must be disposed of only under the terms of an authorisation granted by the Agencies under the Radioactive Substances Act 1993[53] or under an exemption under that Act. Those disposing of such waste should seek the advice of the Radiation Protection Adviser appointed under the Ionising Radiations Regulations 1985,[54] both as regards the protection of staff and permitted means of disposal.

146 Radioactive waste from diagnosis and intensive radiotherapy has low radioactivity and short half-lives. If the waste is a water-miscible fluid, and the discharge authorisation permits, it may be disposed of to the sewer. Consent needs to be obtained from the sewerage undertaker (England and Wales) and the water authority (Scotland).

Incineration

147 Modern clinical waste incinerators can usually deal with a wide range of clinical waste, including cytotoxic drugs. The combustion conditions in modern clinical waste incinerators include:

- high temperature;
- long residence times; and
- controlled temperature change to avoid the formation or reformation of dioxins (and related compounds) in the flue gases.

Such plant has specially adapted handling systems; the need for minimising manual handling is often contained in the authorisation.

148 Standards for incineration processes rated at less than 1 tonne per hour are set out in Department of the Environment Process Guidance Note PG5/1.[55] These processes are regulated by local authorities under Local Authority Air Pollution Control (LAAPC), or in Scotland by SEPA. The authorisations for these incinerators cover only their flue emissions. The operator needs a waste management licence from the waste regulatory body to cover waste handling at the incinerator.[53]

149 Incinerators rated at 1 tonne per hour or more of clinical waste are subject to Integrated Pollution Control (IPC).[55] The responsibility for authorising these incinerators rests with the Agencies.

150 Municipal incinerators are authorised to accept household and commercial wastes. However, suitably designed and run municipal incinerators may be licensed to deal with certain types of low-risk non-infectious waste, including Group E waste, sanpro, treated and/or sterilised wastes. Suitable municipal incinerators are likely to have handling facilities that allow intact packages of healthcare waste to be fed in.

151 Municipal incinerators are normally much larger than clinical waste incinerators. As a result they tend to achieve greater economies of scale, use less energy per unit of throughput, and achieve greater efficiency in energy recovery than dedicated clinical waste incinerators. However, if consignors send clinical waste to a municipal incinerator they need to satisfy themselves under their duty of care that the incinerator is authorised/licensed to accept the waste, and the plant is being managed in accordance with the requirements of the authorisation/licence.

152 In deciding whether to send some clinical waste to a municipal incinerator, the consignor will probably wish to evaluate benefits and costs. It is likely that additional arrangements for sorting the waste will need to be made, if a municipal incinerator is used to treat some of it. The extra cost of sorting may be offset against the municipal incinerator's lower price per tonne. Consignors may also wish to consider whether the environmental benefits - particularly those attributable to energy efficiency and energy recovery - are greater at the municipal or clinical waste incinerator.

Landfill

153 Under the Common Position adopted on the EC draft Landfill Directive (File no. 97/0085(SYN)) dated 3 April 1998, hospital and other clinical wastes arising from medical or veterinary establishments, which are infectious, as defined in the Hazardous Waste Directive 91/689/EEC (property H9 in Annex III), and waste falling within category 14 of Annex I.A of the Directive, cannot be accepted for landfill.

Groups A, B and C wastes

154 Wastes in Groups A, B, and C, must be treated and made safe before they are landfilled, at a licensed site. Making the waste safe may sometimes not be practical, for example waste which is likely to contain infective quantities of a pathogen that:

- continues to be active after treatment;
- persists for long periods outside the human body; and
- can be ingested when wastes are crushed at landfill.

Such waste should not be sent for landfill.

Group D wastes

155 Group D wastes should not be landfilled, but should be incinerated or otherwise treated by a process which destroys the component chemicals. Cytotoxic drugs in particular, must never be landfilled. Site licences explicitly exclude these Group D wastes.

Group E and sanpro wastes

156 Waste in Group E and sanpro can be landfilled at sites licensed to receive it. Landfilling of Group E waste at a licensed site satisfies environmental and human health considerations. Waste managers may properly consider it among the options available to them for Group E waste.

Human tissue

*Placentas may be given to the mother on request.

157 Identifiable human tissue, eg theatre waste, limbs and placentas,* must not be landfilled. No circumstances are sufficiently compelling to justify the relaxation of this rule. Environmental law requires the offensiveness of waste to be taken into account in disposal. Members of the public must not be able to identify human tissue that has been discarded as waste.

158 The only treatment method for identifiable human tissue waste is incineration in a clinical waste incinerator. Municipal waste incinerators are normally not suitable for human tissue waste; neither the waste handling arrangements nor combustion conditions are adequate.

Incinerator ash

159 Incinerator ash should be disposed of at a licensed landfill site.

Summary

160 Table 3 summarises the treament/disposal options suitable for the different groups of clinical waste.

Table 3 *Treatment/disposal option*

Category of waste	Treatment/disposal option
Group A	Clinical waste incinerator.
Group A, other than: 1 identifiable human tissue; 2 any wastes known or likely to contain ACDP Hazard Group 4 biological agents; 3 all waste from containment laboratories level 3; 4 all cultures of ACDP Hazard Group 2 biological agents. *Items 2, 3 and 4 above should be autoclaved before disposal (see paragraph 84)*	Maceration and heat or chemical treatment. The residues may either be incinerated or sent to landfill. NB where these treatments accept waste known or likely to contain ACDP Hazard Group 3 biological agents, then the risk to employees from these agents must be adequately assessed and controlled.
Group B	Clinical waste incinerator.
Group B, other than: 1 any wastes known or likely to contain ACDP Hazard Group 4 biological agents; 2 all waste from containment level 3 laboratories; 3 all cultures of ACDP Hazard Group 2 biological agents. *Items 1, 2 and 3 above should be autoclaved before disposal (see paragraph 84)*	Maceration and heat or chemical treatment. The residues may either be incinerated or sent to landfill. NB where these treatments accept waste known or likely to contain ACDP Hazard Group 3 biological agents, then the risk to employees from these agents must be adequately assessed and controlled.
Un-autoclaved Group C	Clinical waste incinerator.
Autoclaved Group C	Maceration and heat or chemical treatment. The residues may either be incinerated or sent to landfill. Direct to landfill.

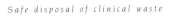
Safe disposal of clinical waste

Group D	Clinical waste incinerator.
Group E	Clinical waste incinerator, municipal waste incinerator,* maceration and heat or chemical treatment. The residues may either be incinerated or sent to landfill. Direct to landfill.
Macerated Group E – excreta and the like	Sewer.
Sanpro	Clinical waste incinerator, municipal waste incinerator, maceration and heat or chemical treatment. The residues may either be incinerated or sent to landfill. Direct to landfill.

* *A municipal waste incinerator specifically designed, constructed, authorised and operated for the incineration of clinical waste may, if so authorised, accept waste from Groups other than E.*

Additional advice for producers other than hospitals

161 Earlier sections of this guidance, on the principles of clinical waste management and the necessary practical precautions, are applicable to small as well as large-scale producers of clinical waste. Experience has shown that the practical precautions taken by smaller scale producers, for example GPs, dental practitioners, nursing homes, residential care homes and special schools, may fail because of a fundamental failure to manage the risk. The following guidance is particularly relevant to small-scale producers.

162 The safe disposal of waste is the responsibility of the independent self-employed practitioner, nursing home or residential care home owner or other employer. The domestic waste collection service should not be used for clinical waste. The options for disposal include:

- local authority special collection and disposal service for clinical waste;
- independent contractor to the local hospital disposal facility;
- independent contractor to the local authority's disposal facility;
- independent contractor to the contractor's disposal facility;
- the practitioner, etc, taking waste by arrangement to the local hospital disposal facility.

163 The categorisation of waste outlined in Table 1 should be adopted. In particular, all Group A clinical waste arising from treatment areas should be transported in UN type-approved sacks or other containers.

164 Sharps (Group B) should be placed in an appropriate sharps container as outlined earlier in this document. On no account should soft drink cans, plastic bottles or similar containers be used for the disposal of needles, since these could present serious hazards to staff if they were disposed of in domestic waste.

165 Clinical waste should be clearly labelled in accordance with CDGCPL[27] before it is removed from site for disposal.

Clinical waste from general medical practitioners

166 GPs should have arrangements in place for:

- identifying clinical waste generated on their premises;
- segregating it in accordance with Table 1; and
- ensuring its correct disposal.

The practice policy should be reviewed and updated regularly. This is particularly important if major changes in the type and quantity of waste generated occur, eg if the amount of invasive treatments increases.

Clinical waste arising from home treatment

167 The situation in the home is different from that in hospitals and other healthcare establishments, because the quantity of waste involved is usually very small. Much of it is produced and handled only by the patient or the family, in circumstances outside the scope of the HSW Act.[15] However, a significant number of employees, including community nurses and home dialysis technicians, work in the home and have to deal with clinical waste. Where employees treat patients at home, employers have a duty to ensure that clinical waste generated is disposed of safely. Arrangements for disposal may be made through an employer's own systems, or by special arrangement with the local collection authority.

168 Employers need to ensure that staff have an adequate supply of UN type-approved packagings and sharps boxes for transporting clinical waste. **Waste sacks may only be used until 31 December 2001**. After this date, clinical waste will have to be transported in type-approved rigid packagings. For additional security, the containers used may need to be lockable.

169 This guidance is also applicable to other healthcare staff who may generate clinical waste away from a hospital or clinic, for example school nurses.

Clinical waste from veterinary centres and practices

170 Some wastes from veterinary practices, including animal tissue and evacuations, drugs or medical products, sharps, swabs, dressings and similar substances, fall within the definition of clinical waste. As elsewhere, the safe disposal of this waste is the responsibility of the employer or self-employed practitioner. Veterinary surgeons are under no statutory obligation to dispose of the bodies of animals, but in many cases will provide this service. Commercial pet crematoria provide disposal service contracts for veterinary surgeons, as do some local authorities.

171 The clinical waste from veterinary practices should not be put into domestic waste. Some local authorities and firms offer a special collection service for veterinary practices. The arrangements for segregation, packaging and storage of the waste before collection should reflect the advice contained elsewhere in this guidance. The waste should be placed in UN type-approved packagings before collection. A designated area suitable for the volume of the waste is required for storage. A refrigerated room or freezer may be necessary if waste is kept for prolonged periods. Sharps generated by the veterinary surgeon on a client's premises should be removed by the veterinary surgeon in a UN type-approved sharps container and taken to the practice for safe disposal.

172 Under the Animal Health Act, special provisions for disposal have to be made in cases of notifiable disease, eg anthrax. This is a responsibility of the Ministry of Agriculture, Fisheries and Food.

Clinical waste arising from research

173 Research may be carried out in dedicated premises or by independent contractors or other employers in premises such as universities, hospitals or clinics. A number of employers may share facilities, and there should be clear agreed arrangements for disposal of clinical waste generated in these areas. It is preferable that the host employer takes the lead on this issue. Full liaison between employers is essential to make sure that the waste is dealt with appropriately and that any special risks are identified, along with safe procedures for dealing with them. These issues need to be specified in the contractual arrangements between the employers. Regulation 9 of the Management Regulations[13] specifically requires employers who share a workplace to co-operate with each other, to ensure that they meet their legal duties under health and safety legislation.

Clinical waste from ambulance services

174 Careful liaison between healthcare providers and ambulance services is essential to ensure that the procedures adopted for waste disposal are compatible. Some ambulances carry pharmaceuticals for use by paramedical staff, and arrangements need to be made for disposal of partly used and outdated stocks.

175 Emergency ambulances need UN type-approved sharps containers of appropriate size, for Category B wastes. The containers should fit into the case used to carry equipment from the ambulance to the patient. They should be securely stored on the ambulance and at the ambulance station. Adequate supplies of UN type-approved sacks or other packagings are needed on the ambulance for other clinical waste. Ambulance services need clear agreed procedures for the removal and disposal of clinical waste and sharps containers. Ways of doing this include local agreements with healthcare providers and central collection systems.

176 Clinical waste may arise during the non-emergency transport of patients, where body fluids or other items contaminate the ambulance. Ambulance services need procedures to deal with this, which include arrangements for the collection, storage and disposal of this waste. Ambulance staff need specific training in cleaning and disinfecting vehicles, in addition to that provided for handling clinical waste. They also need appropriate equipment for cleaning up spillages and suitable personal protective equipment.

Monitoring and review

177 To assess the effectiveness of the arrangements for reducing the risks from clinical waste, employers need systematic procedures for monitoring and review. These procedures need to check in particular that clinical waste is not entering the non-clinical waste stream and that the different categories of clinical waste are treated and disposed of appropriately. Measures which are not working, or which have unforeseen consequences, need to be identified and changed.

178 'Active' monitoring involves checking that systems and procedures are working without waiting until something goes wrong. It is a key part of a successful monitoring regime and part of a line manager's function. Accident and injury data also provide information on the effectiveness of precautions, and helps everyone learn from experience. The introduction of new measures may not reduce the number of reports in the short term. Reviews of incident reporting programmes have often found that reports increase with staff awareness. This is not necessarily a bad thing; it indicates how much the original problem was underestimated, and ought to generate greater confidence in ill-health and accident data.

Investigation of accidents and incidents

179 The Reporting of Injuries, Diseases and Dangerous Occurrences Regulations 1995[56,57] (RIDDOR) require the following to be reported to the appropriate enforcing authority: certain major injuries; all injuries resulting in an employee being unable to carry out their normal duties for more than three days; and specified occupational diseases, including infections reliably attributable to the work activity. For most healthcare premises, the relevant authority is the Health and Safety Executive. Records of these incidents must be kept. Social security legislation requires an accident book or something similar to be kept and accessible to staff. Effective health and safety management systems ensure the internal reporting, recording and investigation of a wider range of accidents and incidents than those which are legally reportable.

180 Incidents such as spillages, damaged packagings, inappropriate segregation or any incident involving sharps, need to be reported to the line manager or other suitable individual, and investigated by them. The investigation of these accidents and incidents needs to establish the cause, and the action needed to prevent a recurrence.

181 The analysis and investigation of incidents involving clinical waste, whether reportable or not, helps identify causes, trends, the level of compliance with current legislation, the effectiveness of the precautions in place, and problem areas for which satisfactory precautions have yet to be provided. Information on issues such as the costs of such incidents is also relevant to senior managers.

182 The depth of each investigation will vary, depending on the nature of the incident. To be worthwhile, however, any investigation needs to consider carefully the underlying causes. Action after an accident will not be effective if the investigation addresses only the superficial and obvious causes, and misses more significant issues.

Monitoring results

183 The active and reactive monitoring of clinical waste procedures is most effective as part of an overall system of health and safety monitoring, with information passing up the line management chain to senior management.

Safe disposal of clinical waste

APPENDIX 1

Classification of and packaging requirements for clinical waste

Waste group	Type of clinical waste	Classification for carriage by road	Packaging where waste is for final disposal on-site	Packaging where waste is transported for final disposal off-site
Group A	Identifiable human tissue, blood, animal carcasses and tissue from veterinary centres, hospitals or laboratories. Soiled surgical dressings, swabs and other similar soiled waste. Other waste materials, from infectious disease cases, excluding any in Groups B-E.	UN 3291, unless the pathogens it contains are known, in which case UN 2814 or UN 2900 should be used as appropriate.	In suitable waste containers.	In clinical waste containers which are type approved under the requirements of CDGCPL2 for waste classified as UN 3291.
Group B	Discarded syringe needles, cartridges, broken glass and any other contaminated disposable sharp instruments or items.	UN 3291, unless the pathogens it contains are known, in which case UN 2814 or UN 2900 should be used as appropriate.	In sharps containers which meet specifications for BS 7320 or equivalent.	In sharps containers which are type approved under the requirements of CDGCPL2 for waste classified as UN 3291.

...cont

Group C	Microbiological cultures and potentially infected waste from pathology departments and other clinical or research laboratories.	In most cases this will have already been sterilised in accordance with HSAC guidance, in which case it is not dangerous for carriage as an infectious substance. In many of the remaining cases, the pathogens will be known, in which case it should be classified as UN 2900 or UN 2814 as appropriate. However, when they are not known, UN 3291 should be used.	In containers suitable to the level of risk.	If pathogen can be specified, the waste will need to be packaged as required for waste classified as UN 2814 or UN 2900. If pathogen cannot be specified, the waste will need to be packaged as required for waste classified as UN 3291.
Group D	Drugs or other pharmaceutical products.	UN 3291 is inappropriate for waste chemicals and medicines which do not contain infectious substances. Where appropriate they should be classified and then packaged in accordance with their properties, just as any other chemical. It is recommended that specialist advice is sought, eg from the pharmacy, on the correct disposal method.	In containers suitable to the level of risk. Clinical waste containers are not suitable for this purpose.	Where chemical wastes and waste pharma-ceutical products are dangerous for carriage, in type approved containers appropriate to the classification of the waste. Clinical waste containers are not suitable for this purpose.
Group E	Items used to dispose of urine, faeces and other bodily secretions or excretions assessed as not falling within Group A. This includes used disposable bed pans or bed pan liners, incontinence pads, stoma bags, and urine containers.	Not dangerous for carriage.	In suitable waste containers.	This waste is not considered dangerous for carriage. The transport regulations do not apply any specific requirements for packaging. The welfare of employees handling the waste will however need to be considered when deciding how the waste is packaged.

APPENDIX 2

First aid for sharps injuries

IMPORTANT MESSAGE TO ALL STAFF -
Action to be taken following a needlestick/sharps injury.

The Department of Health NHS Executive has produced a guidance document - *Post exposure prophylaxis for health care workers occupationally exposed to HIV.*[47]

All XXX NHS health care workers who sustain a needlestick type injury at work should follow the procedure set out below which now includes the action they should take if they have been exposed to blood or other high-risk body fluids known to be infected with HIV.

In these circumstances, by following the 'Action to be taken' (see 3 'Assess infection risk' (c)) they would be seen promptly by a physician, and the best course of action would then be agreed. This may include taking the post exposure prophylaxis (PEP) which is three different medications to be taken daily for four weeks. Clinical care and counselling would also be instigated and continued for as long as necessary.

The risk of acquiring HIV infection from a needlestick injury is small, and through mucous membrane exposure even smaller, but the risk can be reduced still further if the PEP medication is taken as soon as reasonably possible after exposure.

Please remember to follow guidelines and procedures to avoid any incident which may put you or others at risk.

Action to be taken following a needlestick/sharps injury:

1. First aid - encourage wound to bleed. Do not suck. Wash with soap and water. Dry and apply waterproof dressing.

2. Report incident to line manager or senior staff in department. An accident form will need to be completed.

3. Assess infection risk:
   (a) Unused/clean sharp - definitely no risk of infection. Complete accident form and dispatch. If in doubt seek further advice as step (b).
   (b) Used/dirty sharp - **source unknown** or **known**, also human bite/scratch/mucous membrane splash. Seek professional advice from:
   Occupational Health Department - Monday to Friday 08.30 - 16.30
   Tel ....
   In their absence, contact duty microbiologist on call.

(c) Used/dirty sharp or human bite/scratch/mucous membrane splash - **source known** to be diagnosed as HIV Positive.
Seek **immediate** professional advice from:
Occupational Health Department - Monday to Friday 08.30 - 16.30
Tel ....
In their absence, contact specialist registrar for accident and emergency, or duty microbiologist on call via switchboard.

APPENDIX 3

Example of a local clinical waste disposal procedure

Introduction

This particular procedure was drawn up for an acute trust. The principles that it contains may be adapted for all healthcare premises.

Clinical waste is produced in wards and most departments across the Trust. As part of the trust waste policy, this procedure sets out the method to ensure that clinical waste is handled and disposed of safely in accordance with Health and Safety Executive, Environment Agency and best practice requirements.

Definitions

The waste control officer	Site Services Manager.
The producer	**Any person** handling waste before placing it in the secure wheeled container at the waste collection point.
The carrier	XXXX Ltd.
The disposer	YYYY Ltd.
Waste collection point	An area designated to store large wheeled containers for the reception of clinical waste bags and sharps containers.
Wheeled container	Yellow 820 litre, lockable wheeled container used for receiving clinical waste. Can be found at waste collection points.
Clinical waste container	The receptacle used to hold the yellow bags locally in the ward or department.

Roles and responsibilities

The waste control officer

Responsible for ensuring that the Trust manages clinical waste disposal in accordance with its waste management policy, and for the revision of this procedure.

General managers

Responsible for ensuring clinical waste is managed in accordance with this procedure within their area of operational responsibility, and liaising with the waste control officer with regard to all matters arising from the application of this procedure.

Producers

In line with this procedure, ensure:

- clinical waste is segregated and placed into the correct clinical waste container;
- the correct specification of bag is used;
- bags are identified;
- bags are sealed correctly;
- yellow bags and sharps containers are transferred to the waste collection point, and stored safely and securely.

The general porter - refuse

In line with this procedure, ensure:

- transportation of full and empty wheeled containers between waste collection points and storage areas;
- faulty containers are identified and corrective action taken.

The carrier - XXXX Ltd

Ensure the waste is collected from the main storage area and transported to the agreed incinerators as per agreed contract.

The disposer - YYYY Ltd

Ensure the waste collected is disposed of in accordance with the agreed contract.

Safe disposal of clinical waste

What is clinical waste?

Definition

Any waste which consists wholly or partly of human or animal tissue, blood or other body fluids, excretions, swabs or dressings, syringes, needles or other sharp instruments, being waste which, unless rendered safe, may prove hazardous to any person coming into contact with it; and

any other waste arising from medical, nursing, dental, veterinary, pharmaceutical or similar practice, investigation, treatment, care, teaching or research, or the collection of blood for transfusion, **being waste which may cause infection to any person coming into contact with it**.

General

Clinical waste is:	Clinical waste is not:
✓ Human tissue	✗ Waste that is not described in the list opposite
✓ Any items which are or may be soiled by blood or other body fluids, for example: - wound dressings - swabs - disposable gloves and aprons - materials used to clean up spillages (excluding cytotoxic spillages*)	✗ Radioactive waste - see Trust radiation policy
	✗ Flowers
	✗ Newspapers
	✗ Packaging
✓ Colostomy and urine bags	✗ Office waste
✓ Incontinence pads	✗ Kitchen waste
✓ Vomit bowls and sputum pots	✗ Apple cores
✓ Empty IV bags and administration sets	✗ Cans
✓ Sharps - See Trust sharps policy - **do not place sharps in bags**	✗ Bottles
	✗ Aerosols
✓ Waste arising from treatment using cytotoxic drugs*	
✓ Pharmaceutical waste* - return all unused drugs to Pharmacy	

*Refer to cytotoxic waste policy

NB this table is an example of internal guidance which reflects local practice for managing waste within a waste producing unit

APPENDIX 3

Specification of bags/ containers/ enclosures to be used

Clinical waste containers

Shall be capable of containing the waste without spillage. If they are intended for re-use they should be capable of being suitably cleaned and, where required, disinfected. The lid of the container should be capable of being operated without lifting it by hand, ie by foot pedal, and shall close securely.

Clinical waste bags

Waste bags conforming to the appropriate NHS Supplies standard should only be used; these bags having been approved to the relevant UN standard.

Storage areas

Ward/department

Adequate supplies of appropriate clinical waste containers shall be provided local to the area where the waste arises, ie treatment rooms. Waste shall not be allowed to accumulate in corridors or other unsuitable places, nor shall it be stored in such a manner as to cause harm to others.

Waste collection point

Waste collection points shall be established throughout the site. Sufficient numbers of wheeled containers shall be sited at these points to accommodate waste produced locally without overspill. Each container shall be locked, and wards and departments issued a key for their use only.

No waste shall be left at the side or on top of these containers.

Transport

Transport within the hospital

It is the responsibility of the producer to manually remove waste from their area to a waste collection point.

Internal to the building, the general porter - refuse shall manually move wheeled containers to their destinations. Tow trucks shall be used external to the building, to transport full and empty wheeled containers between the collection points and the dedicated storage areas.

Transport off-site

The contractor responsible for the removal of waste shall have a safe system of work in operation.

The waste control officer is responsible for the management of this contract. Portering duty managers and the Estates Department on-call engineer have telephone numbers in case of emergency.

Handling of clinical waste for disposal

- Locate clinical waste container in suitable area.
- Before inserting yellow bag to line the container, affix complete label identifying the ward and department and current date.
- Insert yellow bag to line container.
- Only dispose of wastes as defined in this procedure.
- When the bag is approximately 2/3 full, exchange for empty bag.
- Seal the bag as follows:
  - gather top of the bag;
  - fold the neck of the bag over;
  - tie the neck by forming a loop and passing the end through the loop creating a knot;
  - tighten the knot to ensure an effective seal.
- Ensure label is still attached.
- Carry bag to waste collection point and place into yellow wheeled container ensuring it is locked on completion. If the container is full, take the waste to the next nearest waste collection point, and upon return to your area bleep the general porter - refuse, asking them to exchange the wheeled container in your area. Between 10 pm and 8 am Mon-Fri contact the duty manager on ....

No waste shall be left at the side or on top of these containers.

Training needs

All employees who are required to handle and move waste shall be adequately trained in this procedure. A record of such training shall be kept by each department and maintained up to date. A copy of this procedure shall be made available in all areas that produce clinical waste.

Personal protective equipment (PPE)

When handling clinical waste, suitable gloves should be worn as a minimum.

It shall be the responsibility of the General Manager to ensure that these items are provided, used and maintained.

Accidents and incident reporting and investigation

The Trust's policy on accident and incident reporting should be followed in relation to any incident involving waste. Close liaison between the Waste Management Task Team and Control of Infection Committee shall ensure all incidents are properly investigated to prevent recurrence and appropriate corrective actions taken.

Spillages

When dealing with spillages of blood or body fluids, it is very important to ensure that there is adequate ventilation when using hypochlorite. Any spillage of body fluid must be cleaned up immediately. Wear gloves and a disposable plastic apron. Pour sodium hypochlorite (10 000 ppm) on to paper towels placed over the spill. Wipe up the spill, and then wipe over the disinfected area with sodium hypochlorite (10 000 ppm) and, if possible, leave for five minutes contact time before washing the area with hot water and detergent.

PPE (personal protective equipment) in the form of disposable gloves and apron should be worn.

Complete a Trust accident and incident report form.

Final disposal

All clinical waste shall be disposed of by means of incineration. The waste control officer is responsible for ensuring a suitable contract is established and maintained.

Local rules and information

The key for the yellow wheeled container is held ....
Our waste collection point is located at ....
The next nearest waste collection point is ....

Document control

Approved by	:	Infection Control Committee
		Waste Management Task Team
Date of approval	:	....
Date of next revision	:	....
Responsible for review	:	Waste control officer

APPENDIX 4

Example of a local mercury spillage procedure

Although these procedures were written for an NHS Trust they contain advice which is applicable to other premises where mercury is used.

Introduction

Mercury is a silvery white metal with a bluish tinge, and is liquid at room temperatures, melting at -38 °C. Mercury vapour and all of its compounds are highly toxic. Departments where mercury is used should carry out a risk assessment, as required by the COSHH Regulations. The assessment should not only cover the risks associated with normal use but should also include the risks associated with emergency situations such as spillages. Mercury has a slight vapour pressure even at room temperatures, such that if sufficient of the liquid is exposed in a closed room at normal temperatures, the concentration of the mercury vapour in the air may rise to more than 100 times the current occupational exposure standard of 0.025 milligrams per cubic metre of air.

Mercury may enter the body as a vapour and through the skin. The earliest signs of mercury intoxication include a fine tremor of the fingers and mental changes, a combination of anxiety and aggression known as mercurial erethism. One of the earliest signs is deterioration of handwriting. There is evidence that exposure to low levels of mercury can damage the kidneys.

Use of mercury

Mercury is used mainly in two instruments, thermometers and sphygmomanometers. The major risks in using the instruments lie in dealing with breakage and the resulting spillage of mercury.

There is an opinion that mercury instruments will be less acceptable in the future, and that eventually they will be prohibited due to the risks during manufacture and use. It should be an objective of the Trust to replace, on a rolling programme, all mercury sphygmomanometers in particular.

Mercury spillage procedure

Each area using equipment containing liquid mercury should have available a kit for the current collection and disposal of spilled mercury.

The spillage kit is sited ....

This kit is in a labelled box and consists of a:

- bulb aspirator;
- sealable container containing mercury absorbent paste;
- vapour mask;
- instructions and record sheet;
- plastic shovel;
- brush;
- disposable gloves;
- disposable shoes;
- wooden spatula;
- adhesive tape;
- plastic bag with suitable label.

If a spillage occurs:

1 Segregate area to prevent people walking on the spill and to prevent unnecessary exposure.

2 Wear rubber gloves and mercury vapour mask.

3 Gather as much mercury as possible using brush, plastic shovel, wooden spatula and pipette. Store this mercury under water in the honey jar. Use the adhesive tape to collect as many small droplets as possible, and place this, gloves and all equipment in the bag. Seal the bag with adhesive tape and complete the label with details of spillage and the date.

4 Ensure the area of the spillage is well ventilated and arrange to replace the used items from the spillage kit immediately.

5 Expert advice on dealing with spillage of mercury is available from .... ext. ....

Notes

A **Vacuum cleaners** must never be used to clean spillage, as they spread the mercury vapour widely and will be uncleanable and have to be discarded.

B **Carpets** on which mercury has been spilled must be discarded immediately.

C **Baby incubators** must be withdrawn immediately if a mercury thermometer is broken in them. Expert advice from...........will be required if this happens.

APPENDIX 5
Management checklist

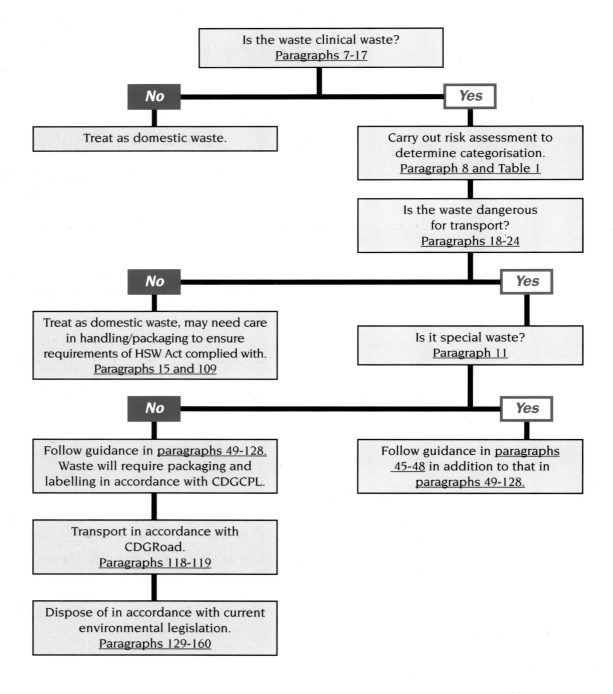

References

1 *The Controlled Waste Regulations 1992* SI 1992/588 HMSO 1992 ISBN 0 11 023588 6

2 *Special Waste Regulations 1996* SI 1996/972 HMSO 1996 ISBN 0 11 062941 8

3 Health and Safety Commission. Advisory Committee on Dangerous Pathogens *Categorisation of biological agents according to hazard and categories of containment* 4th edition HSE Books 1995 ISBN 0 7176 1038 1

4 *Medicines Act 1968* sections 58 and 130 HMSO 1968 ISBN 0 10 546768 5

5 *Healthcare waste: special waste explanatory notes* SWEN001 March 1999. Obtainable from Environment Agency offices, or from the Ecofax fax-back service on 0881 882 288

6 *A guide to the Special Waste Regulations* available from the Scottish Environment Protection Agency

7 *The Carriage of Dangerous Goods (Classification, Packaging and Labelling) and Use of Transportable Pressure Receptacles Regulations 1996* SI 1996/2092 HMSO 1996 ISBN 0 11 062923 X

8 *Approved Carriage List: Information approved for the carriage of dangerous goods by road and rail other than explosives and radioactive material* L90 HSE Books 1999 ISBN 0 7176 1681 9

9 *Approved requirements and test methods for the classification and packaging of dangerous goods for carriage* L88 HSE Books 1996 ISBN 0 7176 1221 X

10 *The management of health and safety in the health services* HSE Books 1994 ISBN 0 7176 0844 1

11 *Control of Substances Hazardous to Health Regulations 1999* SI 1999/437 Stationery Office 1999 ISBN 0 11 082087 8

12 *General COSHH ACOP (Control of substances hazardous to health) and Carcinogens ACOP (Control of carcinogenic substances) and Biological agents ACOP (Control of biological agents). Control of Substances Hazardous to Health Regulations 1999. Approved Codes of Practice* L5 HSE Books 1999 ISBN 0 7176 1670 3

13 *Management of Health and Safety at Work Regulations 1992* SI 1992/2051 HMSO 1992 ISBN 0 11 025051 6

14 *Management of health and safety at work. Management of Health and Safety at Work Regulations 1992. Approved Code of Practice* L21 HSE Books 1992 ISBN 0 7176 0412 8

15 *Health and Safety at Work etc. Act 1974* HMSO 1994 ISBN 0 10 543774 3

16 Health and Safety Commission. Advisory Committee on Dangerous Pathogens *Infection risks to new and expectant mothers in the workplace: a guide for employers* HSE Books 1997 ISBN 0 7176 1360 7

17 *Safety representatives and safety committees. Approved Code of Practice and guidance on the Regulations* 3rd edition L87 HSE Books 1996 ISBN 0 7176 1220 1 (known as 'The Brown Book')

18 *A guide to the Health and Safety (Consultation with Employees) Regulations 1996* L95 HSE Books 1996 ISBN 0 7176 1234 1

19 *The Transport of Dangerous Goods (Safety Advisers) Regulations 1999* SI 1999/257 Stationery Office 1999 ISBN 0 11 080434 1

20 'Council Directive 96/35/EC on the appointment and qualification of safety advisers for the transport of dangerous goods by road, rail and inland waterway' *Official Journal of the European Communities* L145

21 *Are you involved in the carriage of dangerous goods by road or rail* INDG234(rev) HSE Books 1999

22 *The Environmental Protection Act 1990* HMSO 1990 ISBN 0 10 544390 5

23 *The Environmental Protection (Duty of Care) Regulations 1991* SI 1991/2839 HMSO 1991 ISBN 0 11 015853 9

24 Department of the Environment, Scottish Office and Welsh Office *Environmental Protection Act 1990. Section 34. Waste management: the duty of care. A Code of Practice* HMSO 1996 ISBN 0 11 753210 X

25 *Duty of care* from DETR Free Literature, PO Box 236, Wetherby, LS23 7NB reference no. 95EP159

26 *Control of Pollution (Amendment) Act 1989* HMSO 1989 ISBN 0 10 541489 1

27 *Controlled Waste (Registration of Carriers and Seizure of Vehicles) Regulations 1991* SI 1991/1624 HMSO 1991 ISBN 0 11 014624 7

28 *Waste Management Licensing Regulations 1994* SI 1994/1056 HMSO 1994 ISBN 0 11 044 056 0 as amended by the *Waste Management Licensing (Amendment etc) Regulations 1995* SI 1995/288 HMSO 1995 ISBN 0 11 052474 8, the *Waste Management Licensing (Amendment No 2) Regulations 1995* SI 1995/1950 HMSO 1995 ISBN 0 11 053281 3 and the *Waste Management Licensing Regulations 1996* SI 1996/634 HMSO 1995 ISBN 0 11 054324 5

29 *The Water Industry Act 1991* HMSO 1991

30 *Sewerage (Scotland) Act 1968* HMSO 1968

31 *Carriage of Dangerous Goods by Road (Driver Training) Regulations 1996* SI 1996/2094 HMSO 1996 ISBN 0 11 062928 0

32 *Personal Protective Equipment at Work Regulations 1992* SI 1992/2966 HMSO 1992 ISBN 0 11 025832 0

33 *Personal protective equipment at work. Personal Protective Equipment at Work Regulations 1992. Guidance on regulations* L25 HSE Books 1992 ISBN 0 7176 0415 2

34 Medical Devices Agency *Latex sensitisation in the health care setting* Medical Device Bulletin 9601

35 Health Services Advisory Committee *Latex and you* HSE Books (in preparation)

36 Advisory Committee on Dangerous Pathogens *Protection against blood-borne viruses in the workplace: HIV and hepatitis* HMSO 1995 ISBN 0 11 321953 9

37 NHS Estates *Healthcare waste management - segregation of waste streams in clinical areas* Health technical memorandum HTM 2065 Stationery Office 1997 ISBN 0 11 322063 4

38 *Specification for sharps containers* BS 7320:1990

39 Advisory Committee on Dangerous Pathogens *Guidance on precautions for work with humans and animal transmissible spongiform encephalopathies* HMSO 1994 ISBN 0 11 321805 2

40 *Misuse of Drugs Regulations 1973* SI 1973/797 HMSO 1973 ISBN 0 11 030797 6

41 *Destruction of controlled drugs* NHS Executive EL(97)22 available from the Department of Health, Fax 01623 724524, http://www.open.gov.uk/doh/outlook.htm

42 *The Carriage of Dangerous Goods by Road Regulations 1996* SI 1996/2095 HMSO 1996 ISBN 0 11 062926 4

43 *Approved vehicle requirements. Carriage of Dangerous Goods by Road Regulations 1996. Approved requirements* 2nd edition L89 HSE Books 1999 ISBN 0 7176 1680 0

44 *The carriage of dangerous goods explained. Part 1: guidance for consignors of dangerous goods by road and rail* HSG160 HSE Books 1996 ISBN 0 7176 1255 4

45 *The carriage of dangerous goods explained. Part 2: guidance for road vehicle operators and others involved in the carriage of dangerous goods by road* HSG161 HSE Books 1996 ISBN 0 7176 1253 8

46 *First aid at work. Health and Safety (First-Aid) Regulations 1981. Approved Code of Practice and guidance* L74 HSE Books 1997 ISBN 0 7176 1050 0

47 *Guidelines on post-exposure prophylaxis for health care workers occupationally exposed to HIV* Department of Health 1997

48 Public Health Laboratory Service *Chemical disinfection in hospitals* HMSO 1993 ISBN 0 90 114434 7

49 NHS Estates *Clinical waste disposal/treatment technologies (alternatives to incineration)* Health technical memorandum HTM 2075 Stationery Office 1998 ISBN 0 11 322159 2

50 *Trade Effluents (Prescribed Processes and Substances) Regulations 1989* SI 1989/1156 HMSO 1989 ISBN 0 11 097156 6

51 *Water Resources Act 1991* HMSO 1991 ISBN 0 10 545791 4

52 *Water and the environment. Appendix 1* Circular 7/89 Department of the Environment

53 *Radioactive Substances Act 1993* HMSO 1993 ISBN 0 10 541293 7

54 *Ionising Radiations Regulations 1985* SI 1985/1333 HMSO 1985 ISBN 0 11 057333 1

55 Department of the Environment *Chief Inspector's guidance to inspectors. Environmental Protection Act 1990. Waste disposal and recycling: clinical waste incineration* Process guidance note IPR 5/2 HMSO 1992 ISBN 0 11 752652 5

56 *The Reporting of Injuries, Diseases and Dangerous Occurrences Regulations 1995* SI 1995/3163 HMSO 1995 ISBN 0 11 053751 3

57 *The Reporting of Injuries, Diseases and Dangerous Occurrences Regulations 1995: Guidance for employers in the healthcare sector* Health sheet 1 HSE Books 1998

Other reading

Management and disposal of clinical waste Scottish hospital technical note 3 National Health Service in Scotland

Getting sorted: the safe and economic management of hospital waste Audit Commission 1997 ISBN 1 86240 017 2

While every effort has been made to ensure the accuracy of the references listed in this publication, their future availability cannot be guaranteed.

Glossary of terms

ACDP	Advisory Committee on Dangerous Pathogens
The Agencies	Either the Environment Agency for England and Wales or the Scottish Environment Protection Agency for Scotland, or both as the context may require
CDGCPL2	The Carriage of Dangerous Goods (Classification, Packaging and Labelling) and Use of Transportable Pressure Receptacles Regulations 1996
CDGRoad	Carriage of Dangerous Goods by Road Regulations 1996
COSHH	Control of Substances Hazardous to Health Regulations 1999
DETR	Department of the Environment, Transport and the Regions
DOH	Department of Health
DTR2	Carriage of Dangerous Goods by Road (Driver Training) Regulations 1996
EA	Environment Agency
HSAC	Health Services Advisory Committee
HSE	Health and Safety Executive
HSW Act	Health and Safety at Work etc Act 1974
LAAPC	Local Authority Air Pollution Control
NOS	Not otherwise specified
PHLS	Public Health Laboratory Service
IPC	Integrated pollution control
SEPA	Scottish Environment Protection Agency
UN	United Nations

Safe disposal of clinical waste

Printed and published by the Health and Safety Executive   C60   9/99